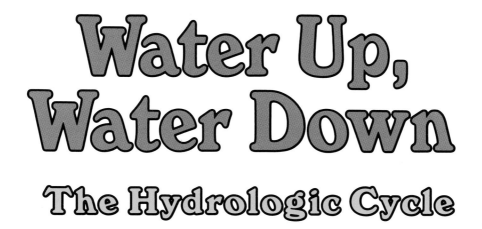

Water Up, Water Down

The Hydrologic Cycle

Water Up, Water Down

The Hydrologic Cycle

For the students at
Braceville,

A Carolrhoda Earth Watch Book

Sally M. Walker

by Sally M. Walker

Carolrhoda Books, Inc./Minneapolis

Carolrhoda Books, Inc. c/o The Lerner Group
241 First Avenue North, Minneapolis, MN 55401

LIBRARY OF CONGRESS CATALOGING-IN-PUBLICATION DATA

Walker, Sally M.
   Water up, water down : the hydrologic cycle / by Sally M. Walker.
      p.   cm.
   "A Carolrhoda earth watch book."
   Includes index.
   Summary: Describes the hydrologic cycle and its importance to life on Earth.
   ISBN 0-87614-695-7
   1. Hydrologic cycle—Juvenile literature. [1. Hydrologic cycle.
2. Water.] I. Title.
GB848.W35   1992
551.48—dc20

91-26144
CIP
AC

Photographs courtesy of: cover, p. 23, Bruce Berg/Visuals Unlimited; pp. 2-3, Dick Poe/Visuals Unlimited; pp. 4-5, Adam Jones; p. 6, Steven C. Wilson; pp. 8, 9 (left), IPS; pp. 9 (right), 33, 45, Steven P. Foley; pp. 10, 27, 28, Patrick Cone; p. 11, S. McCutcheon; p. 12 (bottom), David H. Ellis/Visuals Unlimited; pp. 12-13, 19, David P. Crisman; pp. 14, 24, 29 (left), Sally M. Walker; p. 16 (left, center), Hedberg Aggregates/Jerome Rogers; p. 16 (right), American Association of Petroleum Geologists; p. 17, Calvin Alexander; pp. 20 (left), 38, MPLIC; p. 20 (right), W. Palmer/Visuals Unlimited; p. 21, A.J. Copley/Visuals Unlimited; p. 25, Biblioteca Reale, Turin; p. 29 (right), Tor Eigeland/Aramco World; pp. 30, 40, Ruth Berman; pp. 31, 44, John D. Cunningham/Visuals Unlimited; p. 32, TVA photo by Denise Schmittou; p. 34, Norma Jaxal; p. 35, Jericho Historical Society; p. 36 Dee Culleny/Visuals Unlimited; p. 37, USDA; p. 39, National Center for Atmospheric Research; p. 42, National Association of Conservation Districts. Illustrations by Bryan Liedahl.

Manufactured in the United States of America

2 3 4 5 6 7 8 9 10 – P/JR – 02 00 99 98 97 96 95 94

For my father, who always knew a good story; and my mother and sister, who are always willing to listen.

Thanks to David P. Crisman, P.G., Geological Engineer, for his assistance with this book.

The author wishes to thank James A. Walker for all his help and support; Colin Booth for manuscript critique, once upon a time; Ruth Berman, a wonderful editor who knows how to bring out the best; and Dava Wayman, a special friend whose door is constantly open for my children.

The next time it rains, go outside. Perhaps you'll only stay for a moment, like a cat who doesn't want to get wet. Or maybe you'll linger, letting the rain fall on your hair and dot your skin. Look at the water after it hits the ground. Float a leaf or a stick in a standing puddle or in a curbside stream. When the rain stops, if the air and ground temperatures are just right, maybe you'll be surrounded by mist. Watch as puddles seep into the soil, and notice how your wet skin cools as it dries and the raindrops disappear. Unless it becomes rain or clouds, we can't see water in the air, nor can we see it deep beneath the soil. But it is still there.

Water is always moving between the earth's surface and the atmosphere in a repeated pattern, or cycle. On the earth's surface, water flows across the land and beneath the ground. As the water is heated by the sun, it turns into **vapor,** or gas, and returns to the air—a process called **evaporation.** When it cools, water vapor **condenses,** or turns back into a liquid. It then **precipitates,** or falls to the ground as rain, snow, or ice. The repeated pattern of water movement—flowing on and beneath the earth's surface, evaporating, condensing, and precipitating—is called the hydrologic cycle. The word *hydrologic* comes from two Latin words: *hydr,* which means water, and *logia,* which means study (or knowledge) of a subject.

WHAT IS WATER?

A water **molecule** is made up of two hydrogen **atoms** combined with one oxygen atom. Atoms are tiny particles that form chemical elements. They combine with each other and form molecules. A molecule is the smallest unit of a substance that can exist without changing the substance's chemical composition. Scientists write the chemical formula for water as H_2O. *H* stands for hydrogen, *O* stands for oxygen, and the *2* describes the number of hydrogen atoms. The formula H_2O represents one molecule of water.

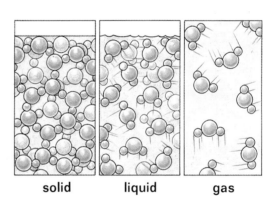

solid   liquid   gas

Water is the only substance on earth that exists in three forms: solid (ice and snow), liquid (water), and gas (water vapor). The form water takes is determined by how tightly its molecules are bound together. When the molecules are tightly bound, there is little or no movement and the molecules form a solid. The bonds in a solid can be broken or bent if enough pressure is applied. Crunching ice with your teeth applies enough force to break molecular bonds. The bonds between molecules in a liquid are much looser, and molecules move around more freely. The bonds can be broken apart fairly easily, which is why water flows and can be poured. Water can be heated to 212° F (100° C), which is its boiling point. Heat energy breaks the bonds between molecules of water, which then escape into the air as vapor. Molecules in vapor move around very quickly and independently, so vapor has no size or shape.

People depend on water for survival. While we are able to live for weeks without food, it's rare for anyone to survive for more than three or four days without water. Each person in the United States uses approximately 65 gallons per day for drinking, bathing, and washing clothes. Industry and agriculture require millions of gallons more. We rely on water for transporting cargo on rivers and oceans. And, of course, we enjoy water when we go swimming, fishing, and boating.

Oceans hold 97 percent of the water on earth. The second largest amount of water, slightly over 2 percent, is frozen in huge, moving bodies of ice, called glaciers. The remaining water, less than 1 percent, is found flowing underneath and on top of the earth's surface. This 1 percent of water, a seemingly small amount, is the source of almost all of the water people use every day.

Above: *Earth as viewed from outer space. Earth is the only planet in our solar system that has a large supply of liquid water at its surface.*
Right: *Water absorbs heat from the sun, keeping temperatures on the earth's surface from becoming too cold or too hot.*

As the sun beats down upon the earth, land surfaces absorb the heat and quickly release it back into the air again. Water, on the other hand, has a larger capacity than land for absorbing heat. Heat is not released into the air as quickly, which helps to keep the earth's atmosphere from becoming too hot. Many of the other planets in our solar system have wide ranges of temperature—at times varying by hundreds of degrees. Water in the earth's atmosphere and on its surface helps prevent this from happening on our planet.

Surface water—water in oceans, rivers, and lakes—is an easily observed part of the hydrologic cycle. We can see surface water as it flows across the land. In fact, surface water often changes the land around us.

Flowing surface water cuts into the land and erodes, or wears away, rock and soil. Pieces of eroded rock and soil, now called **sediment,** may be carried away by the water and deposited elsewhere, forming new land. Water carries heavy sediment by pushing and rolling it along ocean, lake, and river bottoms. Lighter sediment is carried suspended in the water. Sometimes so much sediment is suspended that the water appears cloudy. Water also moves sediment by dissolving it completely.

This waterfall in Utah looks muddy because of the sediment that it is carrying.

Powerful ocean currents carry sediment to new places.

Each year oceans move billions of tons of sediment. Waves batter rocks and sand, changing cliff faces and beaches. Where there are significant changes in seasons, closely spaced, choppy winter waves crash onto beaches, narrowing them by pulling sand back into the ocean. The more widely spaced, lower waves of summer carry sand toward the shore, causing beaches to widen.

Powerful ocean currents carry sediment to new places, in turn building new land areas as the sediment is deposited. Many ocean-side resorts and vacation homes are built on offshore ridges of sand known as barrier islands. Although these islands seem to be secure land, waves generated by powerful hurricanes can be strong enough to wash away buildings and sand.

Above: *The Grand Canyon*
Left: *Over millions of years, the Colorado River
has cut a channel through the Grand
Canyon, dramatically changing the
landscape.*

Grand Canyon National Park, in Arizona, is a dramatic example of how rivers can change a landscape. All rivers flow in a downhill direction in scooped-out paths called channels. About nine million years ago, the land where Grand Canyon National Park is today was a relatively flat area called the Colorado Plateau. The Colorado River used to flow across the plateau. Gradually, certain processes deep inside the earth caused the Colorado Plateau to start rising. The rising land made the channel of the Colorado River steeper, causing the river to flow faster. As the land rose higher, the river flowed even faster, cutting its channel deeper and deeper, until the channel reached its present depth of about 1 mile (1.6 km) in Grand Canyon National Park. Scientists estimate that the river cut the canyon at an average rate of 6½ inches (16.5 cm) every thousand years. From measurements and samples of river water taken in the Grand Canyon, scientists calculate that the Colorado River moves about half a million tons of sediment past any given point each day!

A grassy levee (upper right) *made by people keeps floods from overflowing its bank.*
Note the flooding at left, where there is no levee.

Every time rivers overflow, they carry sediment. When flooding
ends and rivers return to their normal channels, sediment is left
behind. Over time and repeated floodings, some sediment deposits
may form hills along the riverbanks. These hills, called **levees,** are
natural barriers that can help prevent some future flooding.

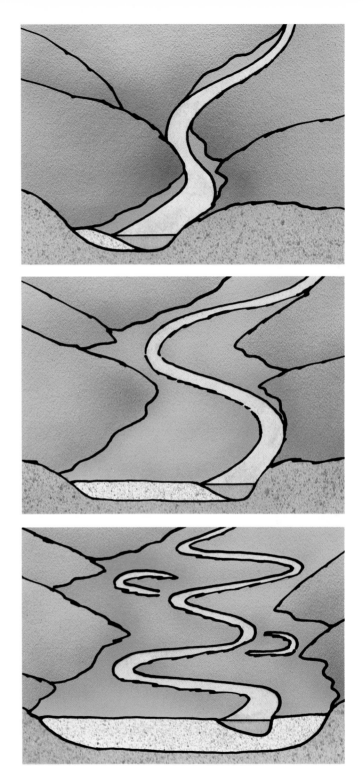

Floodplains (bright green areas) are formed over the years as rivers periodically flood and deposit sediment. Top: *Young River* Middle: *Mature River* Bottom: *Old River*

Flat land alongside rivers is frequently covered by water when flooding occurs. These flat areas, called **floodplains,** are often covered with widespread layers of sediment that have been deposited by past floodwaters. Floodplain sediment can be many inches or feet thick. As a matter of fact, much of the damage done inside homes that have been built on floodplains is caused by thick, muddy sludge left behind as rivers flow back into their channels.

Flooding is normal in river systems. Some rivers flood on a regular basis. Before Egypt's Aswan Dam was built, the Nile River flooded every year and the people counted on the rich sediment deposited by the river for growing their crops. Other rivers may have large floods in 10-year or 50-year periods. From past records and by careful measurements, scientists can predict approximately how often flood conditions might occur, but they cannot predict the exact years.

Many little pores are between individual grains of sand (left). *Larger pores are between the rock pieces of gravel* (middle), *allowing water to flow more easily through gravel.* Right: *A section of porous rock, photographed at 46 times its actual size. A bluish green stain has been applied to show off the pore spaces of the rock.*

The hydrologic cycle continues as water seeps down into the soil and rock beneath the earth's surface. Water flowing below the surface is called **groundwater.** Gravity, the force that prevents us from floating up into the air, pulls water downward through air spaces and cracks in rock and soil. This area is called the **zone of aeration** because of its many air spaces, or pores.

Water seeps into the ground at varying rates depending on the soil. Sandy soil has many little pores between individual grains of sand. Gravel has larger pores between its rock pieces. The larger the pores and passageways, the easier it is for water to trickle through. If a bucket of water is poured onto a pebbly driveway, the water disappears faster than it would in sand because water flows more rapidly through the pebbles' larger pore spaces. Other soil and rocks have smaller air spaces between their grains. Clay, for example, has very tiny grains and pores; water does not flow as rapidly through the tiny air passageways. That's one of the reasons people make pots and bowls out of clay.

Rock that has many air spaces is called **porous rock.** Water often flows inside this kind of rock. When water moves easily through porous rock, the rock is said to be **permeable.**

Usually, groundwater that flows inside porous rock does not move very quickly, especially if the pores and pathways are tiny. Most groundwater flows an average of an inch or less per day, but water can flow as quickly as tens of yards per day in permeable rock. Groundwater follows the easiest route, usually to a lake or stream, but may take thousands of years to reach the land's surface. To find out where groundwater travels and how quickly it is moving, scientists sometimes inject dye into groundwater. They measure how long the coloring takes to reach a nearby well.

Below the zone of aeration, downward-moving water mixes with groundwater that has already seeped beneath the earth's surface. In this area, all the air spaces are already filled with water. Because the passageways between soil and rock grains are saturated, or completely filled with water, this area is called the **zone of saturation.** The top of the zone of saturation is referred to as the **water table.** The water table can be far below the ground or very near the surface. Streams and lakes mark the water table at the land's surface. A shallow hole dug fairly close to a lake's or ocean's shoreline quickly fills with water because the water table is near the surface. In other places, the water table may be many feet beneath the surface. During periods of drought, groundwater drains from the soil into streams, causing the water table to drop even farther below the surface. Sometimes the land surface ends abruptly, perhaps at a valley wall, cliff, or along cracks in rocks. If the water table crosses these places, groundwater flows out. This flowing groundwater is called a spring.

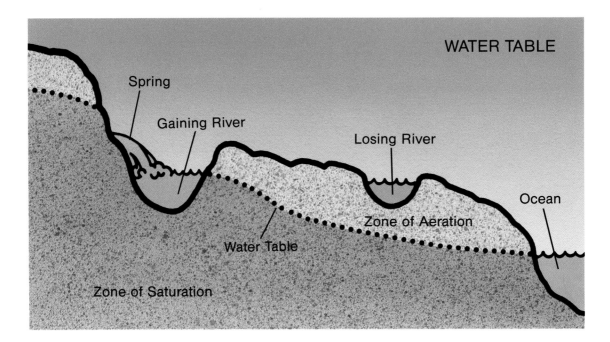

Groundwater flowing through cracks on the face of a cliff is called a spring.

Groundwater is the main source of drinking water for many cities. When people drill wells looking for water, they try to find an **aquifer.** An aquifer is a layer of permeable underground rock saturated with groundwater that can flow easily into wells. Aquifers can be found under more than half the land areas of the United States.

WELLS

Wells are drilled to obtain drinking water. Once an aquifer is located, a powerful drill is used to cut through soil and rock until it reaches below the water table. Usually, strong pumps bring the groundwater to the surface. Once the water reaches the surface, it is pumped into holding reservoirs, huge tanks where water is stored until needed. This water is carefully filtered to remove any bits of soil or rock. Sometimes a chemical called chlorine is added to purify the water and make it completely free of bacteria. Large pipes carry the water to homes and other buildings in cities. Farmers and other people living in the country often have their own small wells to supply them with water.

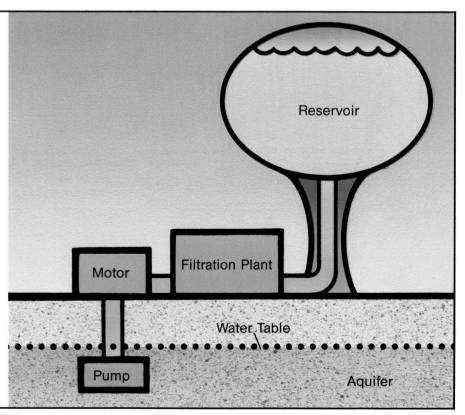

Reservoir

Motor

Filtration Plant

Water Table

Pump

Aquifer

Groundwater, like surface water, can change the shape of the land. As water trickles down through the soil and into the rock below, it may mix with a gas called carbon dioxide, which is found naturally in the air and the soil. When mixed with carbon dioxide, water becomes mildly acidic, which makes it possible to dissolve some types of rock. The solution of water, dissolved rock, and carbon dioxide is carried away to streams and gradually flows to the sea, where it may combine with other solutions and form new rock material. If a hole with an opening to the surface is left where the rock dissolved, the hole is called a cave.

Caves can be many shapes and sizes. Large caves with connecting chambers are known as caverns. The Carlsbad Cavern, in New Mexico, and Mammoth Cave, in Kentucky, are two very large cave groups carved by groundwater.

Caves are cool, dark, and silent places, where the only sound may be the steady dripping of water as it trickles on rock surfaces.

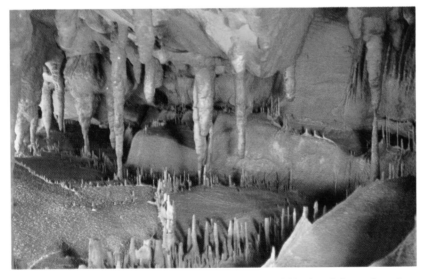

Above: *Mammoth Cave in Kentucky is filled with stalactites forming from the ceiling, stalagmites forming from the ground, and columns.*

Right: *Close-up of a dripping stalactite*

This sinkhole in Florida is partially filled with water and has become a place to live for this alligator.

This water, which frequently carries dissolved rock material, can form marvelous deposits. The most common of these are stalactites. Like icicles made of stone, stalactites are formed when water rich in dissolved rock material drips from cave ceilings. Stalagmites are another deposit formed by water carrying dissolved rock material. They resemble icicles that grow upward from the cave floor. If a stalactite and a stalagmite join, they create a column.

Sinkholes, another land feature created by groundwater, are large pits formed in much the same way as caves are—acidic groundwater dissolves certain rock material. In fact, some sinkholes form when the roofs of caves collapse. And many sinkholes drain into caves.

Sinkholes can be dry or wet. Mexico's Yucatan Peninsula has many sinkholes that are filled with water. The bottoms of these sinkholes cross the water table, so they remain partially filled all the time. For thousands of years, many of these sinkholes have been used as sources of water by people living in the area.

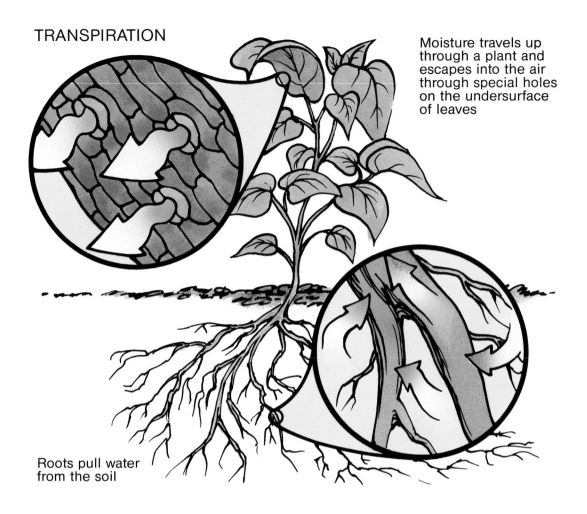

TRANSPIRATION

Moisture travels up through a plant and escapes into the air through special holes on the undersurface of leaves

Roots pull water from the soil

Water found in the soil is not always pulled downward into the rock below. The hydrologic cycle continues as plants use their roots to pull water from the soil. Once inside a plant, the moisture, now called sap, travels through the plant's trunk or stem and out to its leaves. Tiny holes on the undersurface of leaves allow moisture to escape into the air. The process is known as **transpiration.** Water also evaporates from leaf surfaces. During the processes of evaporation and transpiration, a full-grown tree can release as much as 40,000 gallons of water a year—enough to fill a very large swimming pool!

Water also evaporates from oceans, lakes, rivers, and the land's surface, and becomes water vapor in the atmosphere. Heat energy from the sun is used to break the bonds between water molecules, causing evaporation. As molecules are heated, they begin moving rapidly. Scientists say that rapidly moving molecules are "excited." The rapid movement of excited water molecules is strong enough to break their bonds so they can change into vapor.

Water does not evaporate at the same rate everywhere. Warm air absorbs more moisture than cold air. For each 18° F (8° C) increase in air temperature, the air can hold two times more water. Also, warm water evaporates more quickly than cold water. The exact point at which air becomes saturated with water vapor varies according to the temperature.

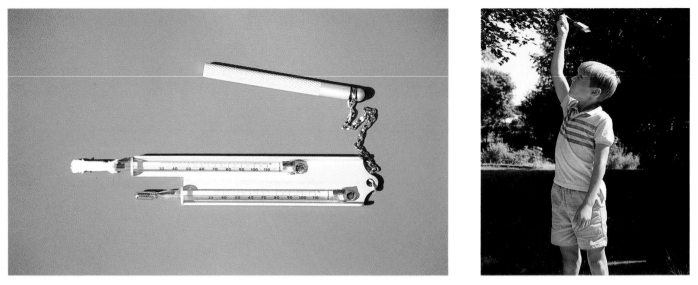

A psychrometer

When a lot of water vapor is in the air, we say the air is **humid.** Scientists frequently use the term **relative humidity,** which describes how much water vapor is in the air at a particular temperature compared with how much water the air at that same temperature is able to hold. An easy way to understand this is to imagine that the air is a towel. If you spill a glass of water, you can wipe up the water with a towel. But the towel probably could absorb more than just a glassful of water. Perhaps it could hold water spilled from 5 or 10 glasses before becoming completely soaked. The amount of water vapor actually present in the air is often only a fraction of the total amount that the air can hold, so relative humidity is expressed as a percentage. When the relative humidity is 100 percent, the air is saturated. Like a towel totally soaked with water, the air can hold no more moisture.

When the relative humidity is 100 percent and the air is saturated, evaporation and precipitation are in a state of balance. As moisture precipitates, the amount of evaporation increases to reach a balanced state again.

MEASURING HUMIDITY

Water vapor in the air is called humidity. Because molecules of water vapor are so small they can't be seen, people who study humidity have developed creative ways to measure the amount of vapor in the air.

Probably the first person to think of an instrument to measure the vapor content of the air was Leonardo da Vinci, a man who was born in Italy in the 15th century. He placed a small wad of dry cotton on one side of a balance scale. Then he placed an object of exactly the same weight as the wad of cotton on the other side of the scale. As the dry cotton absorbed water vapor from the air, it became heavier and the balance pan lowered. The difference between the two weights was the measure of the humidity.

Now scientists use an instrument called a psychrometer to measure relative humidity. A psychrometer is made of two thermometers that are fastened next to one another. The bulb of one thermometer is wrapped in material that is soaked with purified water. To begin measuring relative humidity, a person whirls the psychrometer around until the thermometer with the wet material reaches a steady temperature, which is always lower than the temperature on the dry bulb. The actual air temperature is measured by the thermometer with the dry bulb. The difference between the two temperatures is called the wet-bulb depression and is the result of the evaporation of water from the material. Scientists mark the dry-bulb temperature and the wet-bulb depression on two charts, called Psychrometric Tables, to calculate relative humidity and dew point temperature.

Leonardo da Vinci

Scientists also measure humidity with an instrument called a hair hygrometer. Materials such as wood, cotton, skin, and hair absorb moisture from the air. Human hair gets longer as it absorbs water, increasing its length by about 2½ percent over a relative humidity range of 0 to 100 percent.

On a hair hygrometer, strands of hair are attached to a pointer on a mechanical dial that has been specially marked to indicate relative humidity. The pointer moves as the hair lengthens or shortens. If a written record of relative humidity is needed, a hair hygrometer can be connected to a hygrograph, which has a clock-driven pen that marks a continuous line on graph paper.

APPARENT TEMPERATURE (°F)

Relative Humidity (%)	Air Temperature (°F)										
	70	75	80	85	90	95	100	105	110	115	120
0	64	69	73	78	83	87	91	95	99	103	107
10	65	70	75	80	85	90	95	100	105	111	116
20	66	72	77	82	87	93	99	105	112	120	130
30	67	73	78	84	90	96	104	113	123	135	148
40	68	74	79	86	93	101	110	123	137	151	
50	69	75	81	88	96	107	120	135	150		
60	70	76	82	90	100	114	132	149			
70	70	77	85	93	106	124	144				
80	71	78	86	97	113	136					
90	71	79	88	102	122						
100	72	80	91	108							

When the relative humidity is high, sweat does not evaporate from our skin very quickly, so the air feels warmer to us than it really is. This chart from the National Oceanic and Atmospheric Administration illustrates the difference between real air temperature and apparent air temperature depending on the relative humidity.

The amount of water vapor in the air determines how comfortable we feel on a given day. One way the human body releases heat is by sweating. Evaporation of sweat from our skin helps us cool down on hot days. We feel cooler because during the evaporation process, water molecules require energy to change into vapor. The kind of energy they use is heat energy taken from the water molecules' environment, which in this case is our skin. If the air already contains a lot of water vapor, however, sweat does not evaporate quickly, and we continue to feel hot and uncomfortable.

Molecules of water vapor in the air are so small you cannot see them. For a person to be able to see an object, light must strike the object, which then reflects, or bends, the light back toward a person's eyes. Water molecules are so small that they do not bend back enough light for us to see. As water molecules combine, they form water droplets. For a water droplet to be visible it must contain about 10 billion molecules of water.

Sunlight

Refraction

Reflection

Greatly Magnified Raindrop

Rainbows

Rainbows are unusual enough to make people stop, point, and exclaim. It's almost impossible to keep a smile from your lips when you see one. These magical arches of color stretching across the sky make a rainy day special.

The beautiful colors we see in a rainbow—red, orange, yellow, green, blue, and violet—come about when sunlight shines into a curtain of rain. When light from the sun shines into a raindrop, the light is refracted, or bent, and separated into colors. When it hits the inside back surface of the raindrop, the light is reflected, or turned back. As light leaves the raindrop, it is refracted again, which fans out the colors even more. The angle of the light determines the color we see. Millions of raindrops refracting and reflecting light produce the six-color arch that we call a rainbow.

To see a rainbow you must be facing a rain shower with your back to the sun. If the sky is too cloudy and the sun isn't shining, you will not see a rainbow. Rainbows can also be seen by moonlight, but this is quite rare.

You can create a rainbow of your own with a garden hose. Simply stand with your back to the sun and adjust the spray to a fairly fine mist. Instant rainbow!

When water vapor rises into the atmosphere high above the earth's surface and meets cooler air, it can change in two ways. One way the vapor may change is by turning into water droplets. But if the air temperature is very cold, **deposition**—the change of water vapor directly into solid ice **crystals**—may occur. Deposition also takes place on the ground. Frost is an example of water vapor that has gone through deposition, changing from a vapor into a solid without first becoming liquid water.

Frost

Above: *Dew forms on a plant in a desert.*
Right: *Ice water in the glass cools vapor in the warm air, causing water to condense along the outside of the glass.*

During warm months, the sun heats the air. But the high air temperatures often fall at night after the sun goes down. In the early morning, blades of grass may be coated with water droplets, or **dew,** the result of cooler night air temperatures. (Remember, cool air holds less water than warm air.) When water vapor condenses, dew forms on surfaces. As the air cools, it reaches a point, called the **dew point,** at which it is saturated with water vapor. Condensation begins at the dew point because the air is saturated and cannot hold any more water.

You can do a simple experiment to see how vapor condenses. Fill a drinking glass with cold water and ice cubes. Set it on a table in a warm room or outside on a warm day. The warm vapor in the air will be cooled by the ice water in the glass. After a while, you will notice drops of water condensing on the outside of the glass.

As water vapor high in the atmosphere cools to the dew point, it condenses around dust particles, forming droplets. The droplets come together in larger and larger clumps until they are big enough to reflect light for us to see. These visible clumps of water droplets—and if the air temperature is cold enough, ice crystals—are called clouds.

Millions of water droplets and ice crystals combine to form the large clouds we see in the sky. The cottony clouds that look so fluffy and light actually contain enough water droplets and ice crystals to weigh half a million tons. The enormous black clouds of a thunderstorm may weigh several million tons! When water droplets and ice crystals become too heavy for air currents to hold them suspended in the atmosphere, they fall toward the earth as precipitation.

Wispy cirrus clouds are the highest clouds, with bases higher than 20,000 feet (6,096 m). Clouds that look like cotton puffs have bases that range from 6,500 to 20,000 feet (1,981–6,096 m).

Although clouds can contain enough water to weigh millions of tons, the total amount of water contained in all the clouds at any one time is actually less than .001 of 1 percent of the earth's water supply. If suddenly all the water droplets, ice crystals, and vapor contained in clouds and the atmosphere were to condense and fall evenly over the earth's surface, the total amount would only measure about 1 inch (2.54 cm) of rain.

A cloud that touches the earth's surface and is so thick that visibility is less than .62 mile (1 km) is called fog. If visibility is greater than .62 mile (1 km), the ground-based cloud is called mist.

Fog develops when air becomes chilled to the dew point. This usually occurs when warm, moist air flows over cold water. Fog often forms along the northeastern coast of the United States and Canada. The Gulf Stream—a warm ocean current that flows in a northeasterly pattern from the Gulf of Mexico toward Newfoundland, Canada—brings a steady stream of warm air over the colder North Atlantic Ocean, frequently causing heavy fog to develop. Sometimes the ground temperature helps to cause fog. When warm, moist air passes over colder or snow-covered ground, the cold temperature of the ground lowers the temperature of the air to the dew point, and fog forms.

In 1875 Paul Jean Coulier, a Frenchman, experimented with air and fog. He sealed moist air in a glass container and applied pressure to the air. Fog appeared as the air was squeezed. After repeating the experiment a number of times, fog no longer appeared. Coulier wondered why. Finally, he decided to add new air to the container. He wanted to see if the new air would make a difference. It did. Fog reappeared when he applied pressure to the new air. Water vapor was condensing around something in the new air that was too small to see. Coulier believed that dust particles were attracting water vapor. Dust in the air, much finer than the dust you may find on the furniture in your home, most likely consists of salt particles from ocean spray, volcanic dust, and even molecules of certain gases that have condensed in the air. Coulier concluded that fog appears only when water vapor is able to condense around dust particles that are present in the air. Applying pressure forces vapor to attach itself to dust particles. Once all the dust particles have been used during condensation, no additional fog can appear. The discovery of the role of dust particles in the atmosphere was an important step toward understanding how the hydrologic cycle works.

Snowflakes come in many different shapes and sizes.

Precipitation is another part of the hydrologic cycle. Precipitation begins when water vapor condenses and falls toward the earth as ice, snow, rain, or freezing rain. Ice crystals, tiny hexagonal, or six-sided, crystals found high in the atmosphere, are the first stage in the formation of snowflakes. If an ice crystal continues to grow larger, a snow crystal is formed. Snow crystals, like ice crystals, are six-sided, but they have a more complex shape. A snowflake forms when two or more snow crystals become joined. Some snowflakes may be made of several hundred snow crystals that have come together.

In 1880 Wilson Alwyn Bentley, a 15-year-old boy who lived in Jericho, Vermont, began examining snowflakes through a microscope. He noticed that snowflakes were crystals. Although he was not the first person to notice this, what he did five years later had never been done before and led to a lifelong study of snowflakes. In 1885 Bentley attached a special camera to his microscope and took the first successful photographs of snowflakes. "Snowflake" Bentley, as he came to be called, photographed many thousands of snowflakes during the 40 years that followed, and he never found two that were identical.

Snowflakes come in many sizes. Snow crystals and flakes that form in especially cold air, where there is less water vapor available, tend to be small. Warmer air, with more available moisture, tends to favor the formation of large, wet flakes. These flakes frequently collide with other flakes and stick together while floating downward, sometimes forming snowflakes with diameters as large as 2 inches (5 cm).

Wilson Alwyn Bentley with his special camera

A photograph of a snowflake taken by Bentley

Almost all raindrops begin as snowflakes. They are formed in clouds that are at least partly high enough for the air temperature to remain below freezing. When snowflakes drift into a lower, warmer part of the cloud, the flakes melt and become raindrops.

Like snowflakes, raindrops are not all the same size. They usually range from about .02 inches (.05 cm) to .2 inches (.5 cm) in diameter. The largest raindrops are the ones found during heavy rainstorms, when people are likely to say it's raining cats and dogs. A raindrop size can change as wind tosses it around or as it collides with other raindrops.

Throughout the years 1898 to 1904, Wilson Bentley not only photographed snowflakes, but studied raindrop size as well. He filled several pans with at least 1 inch (2.54 cm) of fine, sifted flour, placed them outside during a storm, and found that each raindrop that landed in the flour-filled pan formed a doughy pellet. After the pellets dried, Bentley measured them—some were almost .25 inch (.64 cm) in diameter. Bentley studied the raindrops from 70 different storms and made 344 raindrop measurements.

During a light rain, about 1,000 raindrops can be found in slightly over a cubic yard of air. During heavy rainstorms, the number of raindrops in the same amount of space can be greater than 5,000.

One of the least known facts about raindrops is that they are not shaped like tear drops. Special wind experiments done in laboratories have made it possible to suspend, or stop, raindrops in midair. Scientists have found that raindrops are actually shaped like hamburger buns!

A heavy rainfall can change the landscape. In moist places, rain falls fairly regularly. Floods in these areas are likely to be widespread and tend to rise and fall slowly, because the soil and plants absorb the rain. Desert areas, however, often have wild, rough floods. Storms in desert areas tend to be heavy rains that last for a short time. Because the soil has been baked dry by the sun, it is difficult for rain to soak in quickly. Since there is little or no vegetation to help slow water runoff and hold soil in place, rain rushes across the land in dirt-filled, swift-moving sheets or streams. This kind of rapidly flowing, dangerous water is called a flash flood. Flash floods usually occur very quickly, almost without warning.

A heavy rainfall in an agricultural area may wash away good topsoil. The faster the water flows, the more topsoil it sweeps away. Topsoil is important because healthy crops depend on this nutrient-rich layer of dirt to grow. Fortunately, farmers can use certain patterns of plowing to slow down flowing water and to reduce the amount of topsoil carried away.

Heavy rain in agricultural areas can wash away nutrient-rich topsoil.

Freezing rain coated this drab-looking tree with sparkling ice.

Freezing rain forms when raindrops are **supercooled,** that is cooled below the freezing point without turning into ice. Upon contact with a cold surface, freezing rain immediately turns from a liquid to a solid, covering everything under a layer of ice. Although the icy layer becomes a sparkling winter spectacle when the sun shines on it, ice-coated tree branches and electrical wires can be dangerous if they become too heavy and snap. Streets and sidewalks covered with a slippery coating of freezing rain are treacherous for traffic and pedestrians.

Another potentially destructive form of precipitation is rounded, lumpy grains of ice, called hail. Balls of hail, usually called hailstones, form during thunderstorms when ice crystals are tossed high inside thunderclouds by strong, turbulent air currents. Supercooled water droplets inside the clouds freeze onto these ice crystals and make each crystal larger by adding new icy layers. This type of growth pattern, called concentric layering, has a small center with layers surrounding it, much like the layering you find when you cut an onion in half. New layers continue to be added until the hailstones become too heavy for the air currents to carry, and the hailstones fall to the ground.

Hailstones range greatly in size. Pea- to marble-sized hailstones are common. During severe storms, air currents can be strong enough to support hailstones the size of golf balls, which can do great damage to plant life and property.

The largest hailstone on record in the United States fell in Coffeyville, Kansas, on September 3, 1970. Its circumference, or the measurement all the way around it, was 17.5 inches (44.45 cm), and it weighed 1.67 pounds (.76 kg).

Several things can happen to precipitation when it reaches the earth's surface. Most precipitation lands on the ground or in surface water. About 15 to 20 percent of the rain that falls on land surfaces soaks into the ground. Some precipitation lands on plants, where it either remains on the leaves and eventually evaporates, or slides off and falls to the ground. More than half of the precipitation is returned to the atmosphere through evaporation and transpiration.

Water can be on the earth's surface and not be an active part of the hydrologic cycle. In or near polar regions or on mountaintops, temperatures are so cold that snow and ice can accumulate in deep layers and eventually form glaciers. Glaciers may last hundreds or even thousands of years. During this time, the frozen water is temporarily removed from the hydrologic cycle. Eventually, however, when the air temperature warms and melting occurs, snow and ice occupy the same place in the hydrologic cycle as rainwater.

A glacier in Chamonix, France.

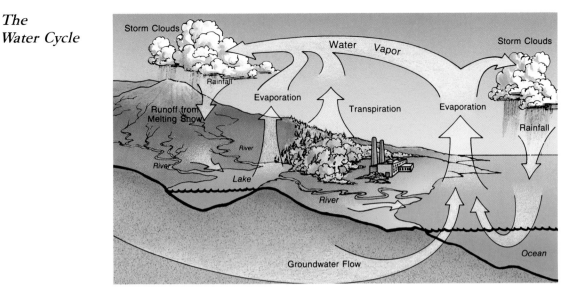

Hydrologists, scientists who study water, have compared the amount of water in oceans with the amount of water on land. The result of their findings is referred to as the water budget. Hydrologists add and subtract the amount of water in much the same way as families add and subtract money in household budgets. After calculating how much precipitation falls on the oceans and how much water evaporates from the oceans, then subtracting evaporation (loss) from precipitation (gain), hydrologists have found that the oceans lose about .36 x 10^{14} cubic meters of water per year. More water evaporates from oceans than precipitates into oceans.

Then hydrologists made the same calculations for land and found that a surplus of about .36 x 10^{14} cubic meters of water falls onto the land per year. The excess amount of water that precipitates onto land is equal to the excess amount of water that evaporates from oceans!

The reason that the land doesn't become soggy and the oceans don't dry up is that the extra precipitation on land ends up in rivers and groundwater. This excess water flows, drips, and seeps its way back to the oceans, where evaporation occurs on a large scale, continuing the hydrologic cycle.

Dangerous chemicals are carelessly thrown into garbage dumps, polluting groundwater sources.

Many people are concerned about the cleanliness of water in all stages of the hydrologic cycle. By nature, groundwater is fairly pure; the tiny pores it flows through remove many impurities. But we hear more and more stories in the news about polluted water. Many things cause water to become dirty and unfit for use. Old tires, cans, bottles, and plastic bags are frequently found in lakes and streams. Dangerous materials in garbage dumps—cans of cleaning fluids, old batteries, gasoline and motor oil, plastic, and styrofoam—can pollute water. This type of pollution is easily seen. Other kinds of pollution are harder to detect.

Chemical fertilizers and pesticides used by farmers and home owners for lawn care are another source of water pollution. Rain leaches, or dissolves and washes away, many of these pollutants, carrying them into the ground below as the water soaks in. These pollutants can contaminate groundwater systems.

Some forms of pollution are fairly easy to remove. Human waste, for example, can be removed by circulating the water through 30 to 40 yards (27.4-36.6 m) of a sand, clay, and organic soil mixture. Some fertilizers and chemicals, however, cannot be removed by circulating water through rock and soil, no matter how small the pores are. Pumping polluted water up into a series of wells and pumping clean water into the ground to replace the dirty water is possible, but it takes many years for the clean water to reach the wells again since groundwater moves so slowly. At this time, our technology cannot completely clean aquifers that have been contaminated. We can only clean the aquifer to a certain point, then natural processes must do the rest—and that could take hundreds of years.

New studies being done by the United States Geological Survey indicate that some weed killers routinely used on corn and soybean crops are polluting rainwater. Several chemicals found in certain weed killers make their way into the atmosphere by evaporation. Once in the atmosphere, the chemicals mingle with water vapor inside clouds. When it rains, the chemicals are mixed with the rainwater. Wind can add to the problem by blowing chemically contaminated clouds to new areas, so the polluted rain falls far from where the chemicals were used.

Conserving Water

Conservation, the protection and careful use of water, is extremely important. In California, an average of 15 billion gallons of groundwater are used each day. More water is used in California than in any other state, partially because of a continuing drought that has affected the area for the past few years. The drought and the demand for water is causing serious problems. In some areas of California, the supply of surface water is so low that the level of the land surface is actually dropping as an increasing amount of groundwater is removed. If this type of water use continues, the aquifers may eventually run out of water.

Fortunately, people are working to teach others how to use water wisely. Some cities in California now have laws that prohibit the use of water for watering lawns. By planting vegetation native to dry areas, people can save millions of gallons of water formerly used on grassy lawns. In an effort to conserve water, a number of communities limit the amount of water people may use each day.

Everyone, not just the people living in drought-stricken areas, should be concerned about water conservation. Simple things—like telling an adult if a faucet is dripping and in need of repair, or keeping a bottle of drinking water in the refrigerator—can save many gallons of water. You can make a difference by not wasting water.

ACID RAIN

An acid is a sharp, bitter-tasting chemical solution. Some strong acids are harsh enough to dissolve certain materials and are harmful to living beings. Other acids, like orange juice or vinegar, are fairly weak and are not harmful to us. Scientists compare acids on the pH scale. Numbers on the pH scale range from 1 to 14: the lower the number, the harsher the acid. Numbers less than 7 are used to compare acids. Numbers greater than 7 are used to compare bases, which are other kinds of corrosive chemical compounds. Pure water has a pH of 7 and is considered neutral; lemons have a pH of 2. Acid rain has a pH of 5.6 or lower. Although most rain is slightly acidic, acid rain may be more than 200 times more acidic than normal rain.

When we burn fossil fuels—coal, oil, and gasoline—several gases are produced. Two of these gases, sulfur dioxide and nitrogen oxide, are released into the atmosphere from car exhausts, power plants, and other manufacturing industries. Once in the atmosphere, these gases combine with water vapor to form sulfuric acid and nitric acid. When droplets of these two acids build up inside clouds and become too heavy to remain in the air, they fall to earth as acid rain.

Acid rain can lower the pH of lakes and streams, making it difficult for fish and other water life to survive. Scientific investigations indicate that acid rain may also react with nutrients and metals normally found in soil, changing them in ways that make it difficult for some tree root systems to grow. The sulfuric acid in rain reacts chemically with certain kinds of building materials—limestone and marble for example—and actually dissolves away bits of the stone.

The Clean Air Act of 1970 restricts the amount of pollutants industry is permitted to release into the atmosphere, so there have been slight improvements. But more restrictions are needed. Manufacturing cars with greater gasoline efficiency would help lower the amount of fossil fuels being burned. The most effective way to reduce acid rain, however, is to decrease the amount of sulfur dioxide and nitrogen oxide produced by power plants, particularly those that burn coal. Older power plants can be refurbished with coal-burning systems designed to reduce the level of sulfur dioxide by at least 80 percent, and nitrogen oxide by more than 50 percent. It is our responsibility to insist that industry be required to do this.

There are ways to prevent groundwater from becoming polluted. Poisonous waste materials can be stored in containers resistant to chemical dissolving, and they should be dumped in areas far from groundwater systems.

The water we now have on earth—on the surface, beneath the ground, and in the atmosphere—is all we will ever have. It is up to us to keep it clean and to use it wisely.

So the next time it rains and you go outside, you'll notice the clouds gliding across the sky. You'll look at the puddles seeping into the soil and the streams of water trickling across the ground. And you'll know that what you are watching is only a small portion of one of earth's important processes: the hydrologic cycle.

GLOSSARY

aquifer: an underground layer of permeable rock filled with groundwater

atoms: tiny particles that are considered to be sources of potential energy

condenses: changes to a denser form, as from vapor to liquid water

crystals: solids with smooth, flat surfaces that are made up of atoms arranged in a pattern

deposition: the act of changing from a vapor directly to a solid without first becoming a liquid

dew: moisture that condenses on surfaces as a result of cool air temperatures

dew point: the temperature at which the air is saturated and a vapor begins to condense

evaporation: the process of liquid water changing to vapor

floodplains: flat areas of land along a river that are at times covered by floodwaters

groundwater: water flowing beneath the earth's surface

humid: the quality of the air when it contains a lot of moisture

levees: ridges of sediment deposited by floodwaters. Some levees are built by people to prevent flooding.

molecule: the smallest unit of a substance, containing at least one atom, that keeps all of the substance's properties

permeable: having pores that allow liquid to flow through

porous rock: rock that contains many air spaces

precipitates: to cause water vapor to condense and fall to the ground as rain, snow, or ice

relative humidity: the comparison of the actual amount of water vapor in the air with the greatest amount possible at that same temperature

sediment: pieces of rock and soil carried and deposited by water, wind, or glaciers

sinkholes: hollow pits or collapsed caverns that may or may not be filled with water

supercooled: cooled below the freezing point without turning to a solid

surface water: water, like oceans, rivers, and lakes, on the earth's surface

transpiration: the process of moisture escaping from a plant through tiny holes, or pores

vapor: a gas; one of the three forms of water

water table: the top part of the portion of the ground that is completely saturated with water; the top of the zone of saturation

zone of aeration: the area of underground rock that has many air spaces in the form of holes and cracks

zone of saturation: the area of underground rock below the zone of aeration that is completely filled with water

Index

ABOUT THE AUTHOR

Sally M. Walker earned her B.A. degree in Physics and Planetary Sciences at Upsala College, in New Jersey. She now lives in Illinois with her husband and two children. She writes during times snatched between volunteer work at her children's school and a thousand and one daily errands. When she writes, Ms. Walker is assisted by her cat, who sits on the desk next to the typewriter, and her golden retriever, who wedges herself between Ms. Walker's feet and takes a nap.